Help Me Understand Genetics
Inheriting Genetic Conditions

Reprinted from https://ghr.nlm.nih.gov/

Lister Hill National Center for Biomedical Communications
U.S. National Library of Medicine
National Institutes of Health
Department of Health & Human Services

Published January 17, 2017

Inheriting Genetic Conditions

Table of Contents

What does it mean if a disorder seems to run in my family?	3
Why is it important to know my family medical history?	6
What are the different ways in which a genetic condition can be inherited?	8
If a genetic disorder runs in my family, what are the chances that my children will have the condition?	20
What are reduced penetrance and variable expressivity?	23
What do geneticists mean by anticipation?	25
What are genomic imprinting and uniparental disomy?	26
Are chromosomal disorders inherited?	28
Why are some genetic conditions more common in particular ethnic groups?	29

What does it mean if a disorder seems to run in my family?

A particular disorder might be described as "running in a family" if more than one person in the family has the condition. Some disorders that affect multiple family members are caused by gene mutations, which can be inherited (passed down from parent to child). Other conditions that appear to run in families are not caused by mutations in single genes. Instead, environmental factors such as dietary habits or a combination of genetic and environmental factors are responsible for these disorders.

It is not always easy to determine whether a condition in a family is inherited. A genetics professional can use a person's family history (a record of health information about a person's immediate and extended family) to help determine whether a disorder has a genetic component. He or she will ask about the health of people from several generations of the family, usually first-, second-, and third-degree relatives.

Degrees of relationship	Degrees of relationship Examples
First-degree relatives	Parents, children, brothers, and sisters
Second-degree relatives	Grandparents, aunts and uncles, nieces and nephews, and grandchildren
Third-degree relatives	First cousin

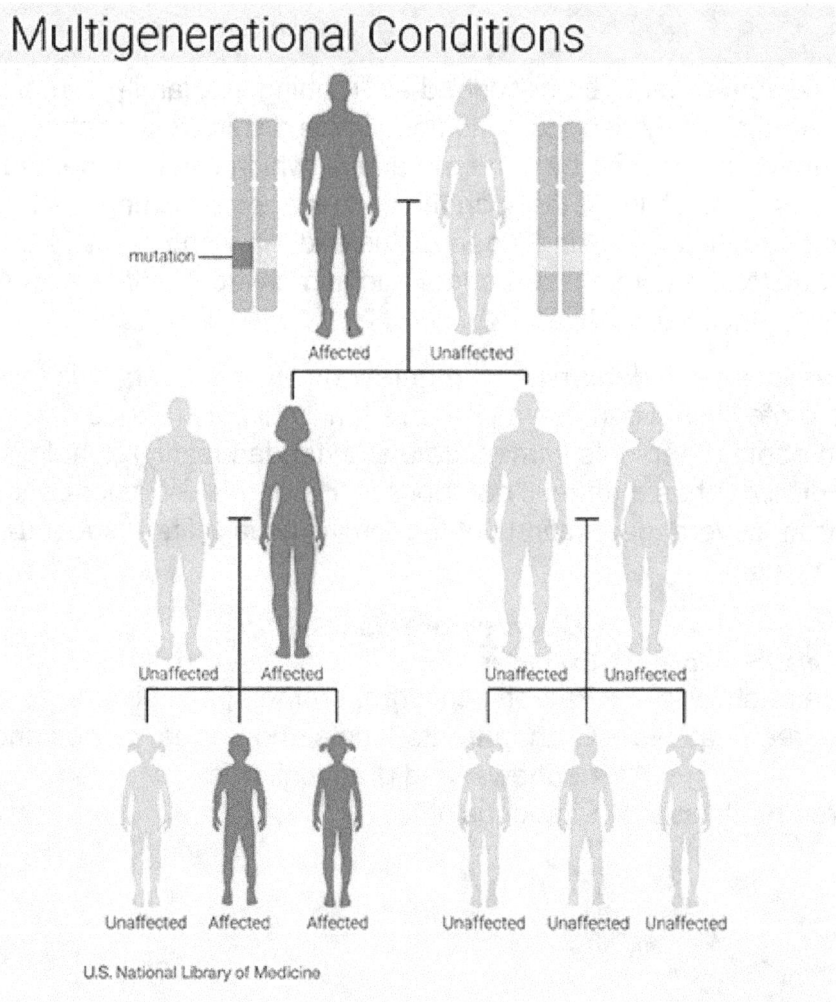

This condition affects members in each generation of a family.

For general information about disorders that run in families:

Genetics Home Reference provides consumer-friendly summaries of genetic conditions (https://ghr.nlm.nih.gov/condition). Each summary includes a brief description of the condition, an explanation of its genetic cause, and information about the condition's frequency and pattern of inheritance.

The Coriell Personalized Medicine Collaborative provides a brief introduction to heritable diseases in the article Heredity: It Runs in the Family (https://cpmc.coriell.org/genetic-education/it-runs-in-the-family).

The Genetic Science Learning Center at the University of Utah offers interactive tools about disorders that run in families (http://learn.genetics.utah.edu/content/history).

The National Human Genome Research Institute offers a brief fact sheet called Frequently Asked Questions About Genetic Disorders (https://www.genome.gov/19016930).

The Centre for Genetics Education provides an overview of genetic conditions (http://www.genetics.edu.au/Publications-and-Resources/Genetics-Fact-Sheets/FactSheetGeneticConditions).

Why is it important to know my family medical history?

A family medical history is a record of health information about a person and his or her close relatives. A complete record includes information from three generations of relatives, including children, brothers and sisters, parents, aunts and uncles, nieces and nephews, grandparents, and cousins.

Families have many factors in common, including their genes, environment, and lifestyle. Together, these factors can give clues to medical conditions that may run in a family. By noticing patterns of disorders among relatives, healthcare professionals can determine whether an individual, other family members, or future generations may be at an increased risk of developing a particular condition.

A family medical history can identify people with a higher-than-usual chance of having common disorders, such as heart disease, high blood pressure, stroke, certain cancers, and diabetes. These complex disorders are influenced by a combination of genetic factors, environmental conditions, and lifestyle choices. A family history also can provide information about the risk of rarer conditions caused by mutations in a single gene, such as cystic fibrosis and sickle cell disease.

While a family medical history provides information about the risk of specific health concerns, having relatives with a medical condition does not mean that an individual will definitely develop that condition. On the other hand, a person with no family history of a disorder may still be at risk of developing that disorder.

Knowing one's family medical history allows a person to take steps to reduce his or her risk. For people at an increased risk of certain cancers, healthcare professionals may recommend more frequent screening (such as mammography or colonoscopy) starting at an earlier age. Healthcare providers may also encourage regular checkups or testing for people with a medical condition that runs in their family. Additionally, lifestyle changes such as adopting a healthier diet, getting regular exercise, and quitting smoking help many people lower their chances of developing heart disease and other common illnesses.

The easiest way to get information about family medical history is to talk to relatives about their health. Have they had any medical problems, and when did they occur? A family gathering could be a good time to discuss these issues. Additionally, obtaining medical records and other documents (such as obituaries and death certificates) can help complete a family medical history. It is important to keep this information up-to-date and to share it with a healthcare professional regularly.

For more information about family medical history:

NIHSeniorHealth, a service of the National Institutes of Health, provides information and tools (http://nihseniorhealth.gov/creatingafamilyhealthhistory/whycreateafamilyhealthhistory/01.html) for documenting family health history. Additional information about family history (https://medlineplus.gov/familyhistory.html) is available from MedlinePlus.

Educational resources related to family health history (https://geneed.nlm.nih.gov/topic_subtopic.php?tid=5&sid=13) are available from GeneEd.

The Centers for Disease Control and Prevention's (CDC) Office of Public Health Genomics provides information about the importance of family medical history (http://www.cdc.gov/genomics/famhistory/index.htm). This resource also includes links to publications, reports, and tools for recording family health information.

The Office of the Surgeon General offers a tool called My Family Health Portrait (https://familyhistory.hhs.gov/) that allows you to enter, print, and update your family health history.

The American Medical Association provides family history tools (http://www.ama-assn.org/ama/pub/physician-resources/medical-science/genetics-molecular-medicine/family-history.page), including questionnaires and forms for collecting medical information.

Links to additional resources (http://www.kumc.edu/gec/pedigree.html) are available from the University of Kansas Medical Center. The Genetic Alliance also offers a list of links to family history resources (http://www.geneticalliance.org/programs/genesinlife/fhh).

What are the different ways in which a genetic condition can be inherited?

Some genetic conditions are caused by mutations in a single gene. These conditions are usually inherited in one of several patterns, depending on the gene involved:

Patterns of inheritance

Inheritance pattern	Description	Examples
Autosomal dominant	One mutated copy of the gene in each cell is sufficient for a person to be affected by an autosomal dominant disorder. In some cases, an affected person inherits the condition from an affected parent (image on page 12). In others, the condition may result from a new mutation in the gene and occur in people with no history of the disorder in their family (image on page 13).	Huntington disease, Marfan syndrome
Autosomal recessive	In autosomal recessive inheritance, both copies of the gene in each cell have mutations (image on page 14). The parents of an individual with an autosomal recessive condition each carry one copy of the mutated gene, but they typically do not show signs and symptoms of the condition. Autosomal recessive disorders are typically not seen in every generation of an affected family.	cystic fibrosis, sickle cell disease

Inheritance pattern	Description	Examples
X-linked dominant	X-linked dominant disorders are caused by mutations in genes on the X chromosome, one of the two sex chromosomes in each cell. In females (who have two X chromosomes), a mutation in one of the two copies of the gene in each cell is sufficient to cause the disorder. In males (who have only one X chromosome), a mutation in the only copy of the gene in each cell causes the disorder (image on page 15). In most cases, males experience more severe symptoms of the disorder than females. A characteristic of X-linked inheritance is that fathers cannot pass X-linked traits to their sons (no male-to-male transmission).	fragile X syndrome
X-linked recessive	X-linked recessive disorders are also caused by mutations in genes on the X chromosome. In males (who have only one X chromosome), one altered copy of the gene in each cell is sufficient to cause the condition. In females (who have two X chromosomes), a mutation would have to occur in both copies of the gene to cause the disorder (image on page 16). Because it is unlikely that females will have two altered copies of this gene, males are affected by X-linked recessive disorders much more frequently than females. A characteristic of X-linked inheritance is that fathers cannot pass X-linked traits to their sons (no male-to-male transmission).	hemophilia, Fabry disease
Y-linked	A condition is considered Y-linked if the mutated gene that causes the disorder is located on the Y chromosome, one of the two sex chromosomes in each of a male's cells. Because only males have a Y chromosome, in Y-linked inheritance, a mutation can only be passed from father to son (image on page 17).	Y chromosome infertility, some cases of Swyer syndrome

Inheritance pattern	Description	Examples
Codominant	In codominant inheritance, two different versions (alleles) of a gene are expressed, and each version makes a slightly different protein (image on page 18). Both alleles influence the genetic trait or determine the characteristics of the genetic condition.	*ABO* blood group, alpha-1 antitrypsin deficiency
Mitochondrial	Mitochondrial inheritance, also known as maternal inheritance, applies to genes in mitochondrial DNA. Mitochondria, which are structures in each cell that convert molecules into energy, each contain a small amount of DNA. Because only egg cells contribute mitochondria to the developing embryo, only females can pass on mitochondrial mutations to their children (image on page 19). Conditions resulting from mutations in mitochondrial DNA can appear in every generation of a family and can affect both males and females, but fathers do not pass these disorders to their daughters or sons.	Leber hereditary optic neuropathy (LHON)

Many health conditions are caused by the combined effects of multiple genes or by interactions between genes and the environment. Such disorders usually do not follow the patterns of inheritance described above. Examples of conditions caused by multiple genes or gene/environment interactions include heart disease, diabetes, schizophrenia, and certain types of cancer. For more information, please see What are complex or multifactorial disorders? (https://ghr.nlm.nih.gov/primer/mutationsanddisorders/complexdisorders)

Disorders caused by changes in the number or structure of chromosomes also do not follow the straightforward patterns of inheritance listed above. To read about how chromosomal conditions occur, please see Are chromosomal disorders inherited? on page 28

Other genetic factors sometimes influence how a disorder is inherited. For an example, please see What are genomic imprinting and uniparental disomy? on page 26

For more information about inheritance patterns:

Resources related to heredity/inheritance patterns (https://geneed.nlm.nih.gov/topic_subtopic.php?tid=5) and Mendelian inheritance (https://geneed.nlm.nih.gov/topic_subtopic.php?tid=5&sid=6) are available from GeneEd.

The Centre for Genetics Education provides information about many of the inheritance patterns outlined above:

- Autosomal dominant inheritance (http://www.genetics.edu.au/Publications-and-Resources/Genetics-Fact-Sheets/FactSheetADInheritance)
- Autosomal recessive inheritance (http://www.genetics.edu.au/Publications-and-Resources/Genetics-Fact-Sheets/FactSheetARInheritance)
- X-linked dominant inheritance (http://www.genetics.edu.au/Publications-and-Resources/Genetics-Fact-Sheets/FactSheetXLDInheritance)
- X-linked recessive inheritance (http://www.genetics.edu.au/Publications-and-Resources/Genetics-Fact-Sheets/FactSheetXLRInheritance)
- Mitochondrial inheritance (http://www.genetics.edu.au/Publications-and-Resources/Genetics-Fact-Sheets/FactSheetMitochondria)

EuroGentest also offers explanations of Mendelian inheritance patterns:

- Autosomal dominant inheritance (http://www.eurogentest.org/index.php?id=614)
- Autosomal recessive inheritance (http://www.eurogentest.org/index.php?id=619)
- X-linked inheritance (http://www.eurogentest.org/index.php?id=623)

Additional information about inheritance patterns is available from The Merck Manual (http://www.merckmanuals.com/professional/special-subjects/general-principles-of-medical-genetics/single-gene-defects).

Images

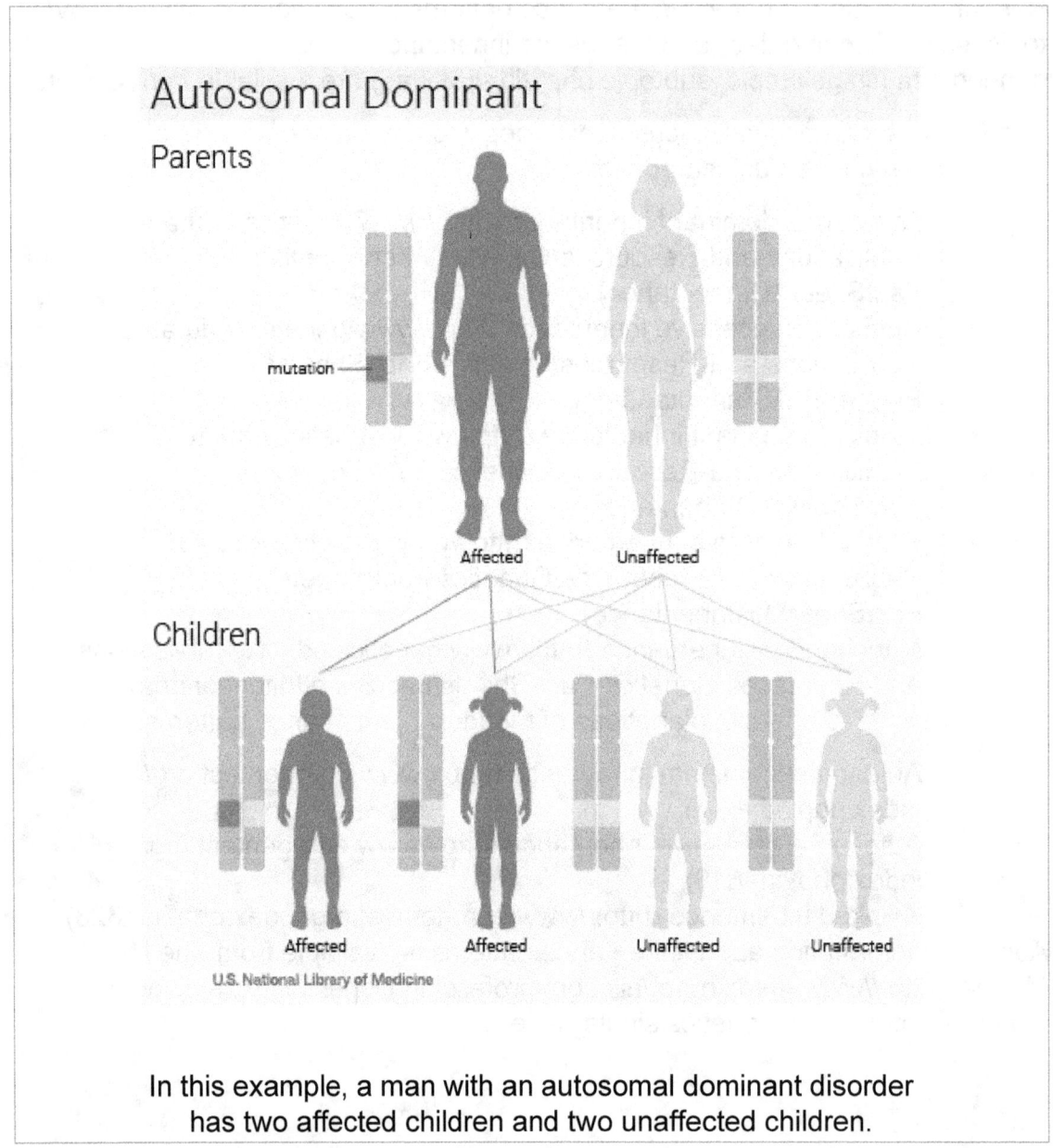

In this example, a man with an autosomal dominant disorder has two affected children and two unaffected children.

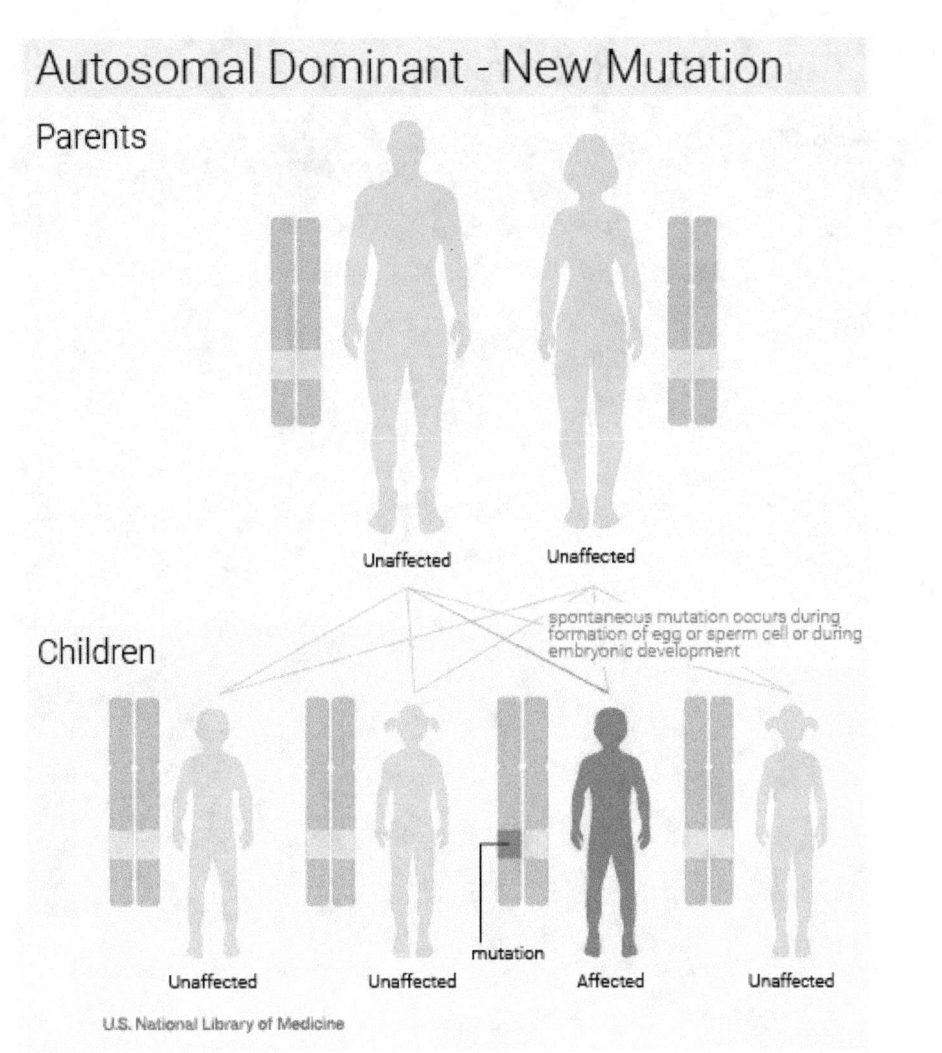

In this example, a child with an autosomal dominant condition has the disorder as a result of a new (de novo) mutation that occurred during the formation of an egg or sperm cell or early in embryonic development.

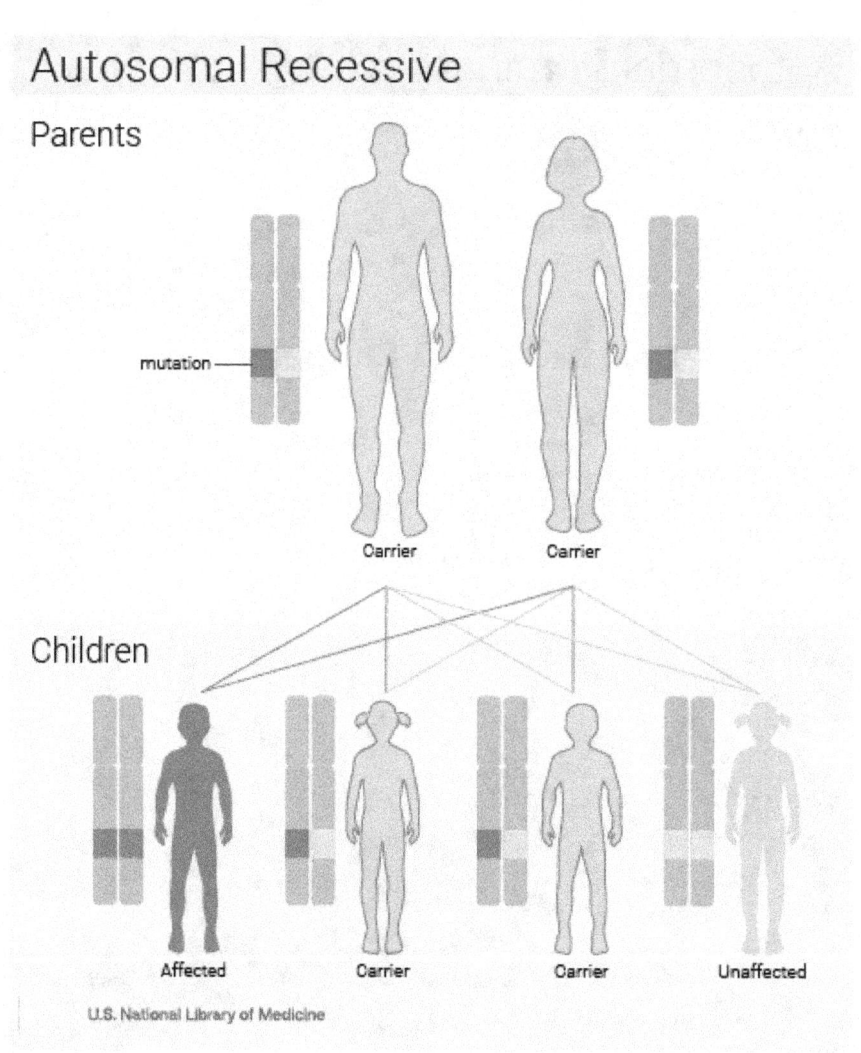

In this example, two unaffected parents each carry one copy of a gene mutation for an autosomal recessive disorder. They have one affected child and three unaffected children, two of which carry one copy of the gene mutation.

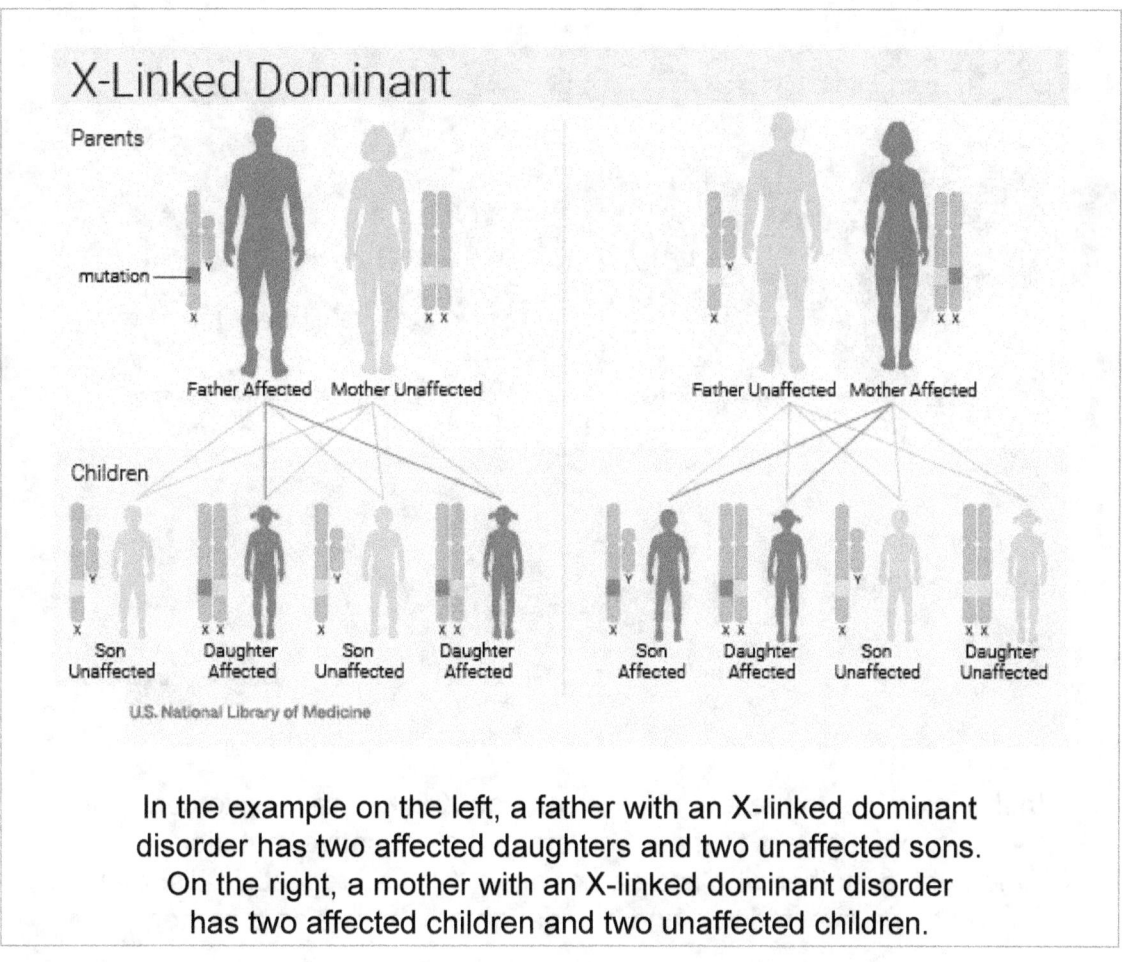

In the example on the left, a father with an X-linked dominant disorder has two affected daughters and two unaffected sons. On the right, a mother with an X-linked dominant disorder has two affected children and two unaffected children.

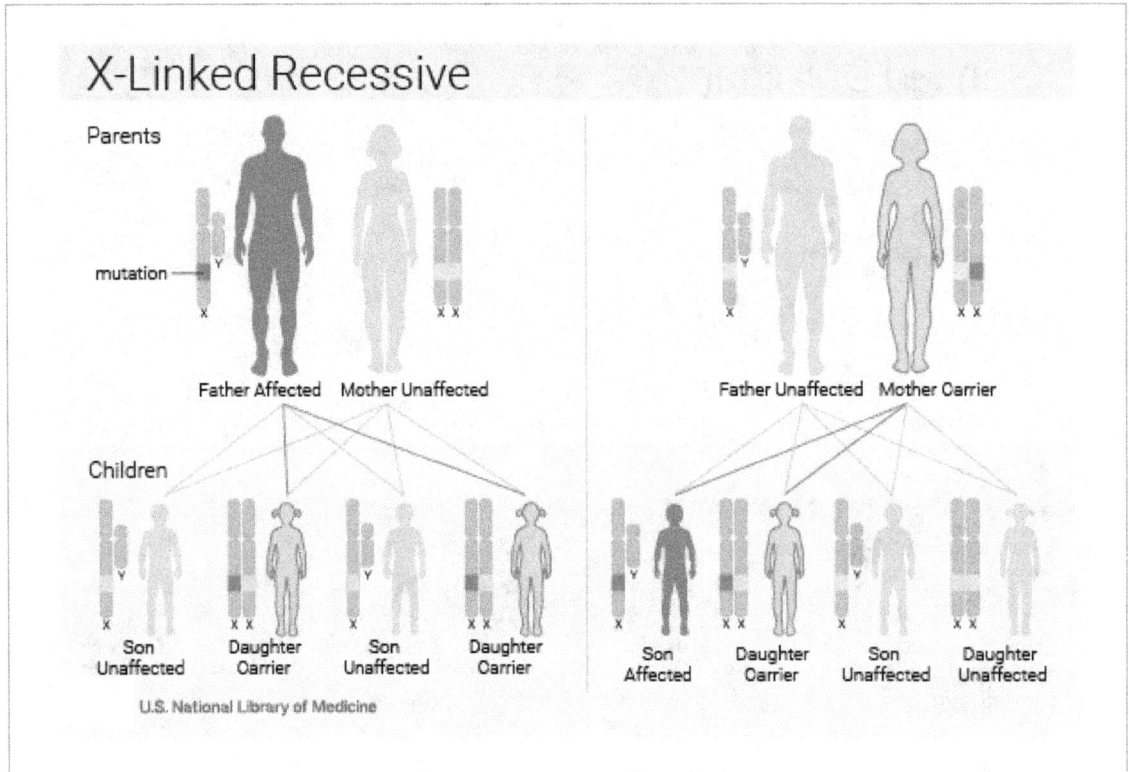

In the example on the left, a father with an X-linked recessive condition has two daughters that are carriers of the causative mutation. On the right, a mother who is a carrier of an X-linked recessive disorder has one affected son and one daughter who is also a carrier.

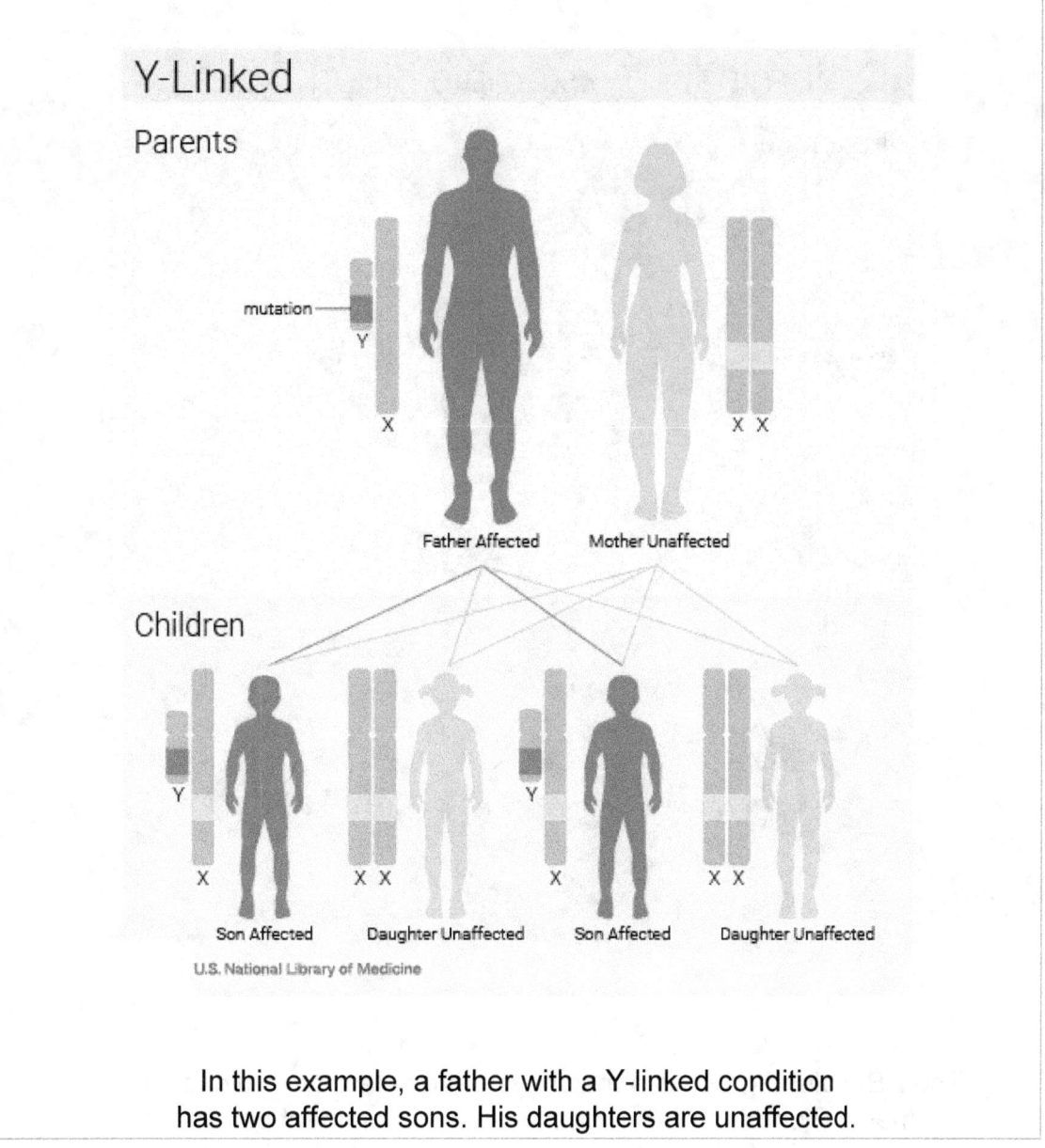

In this example, a father with a Y-linked condition has two affected sons. His daughters are unaffected.

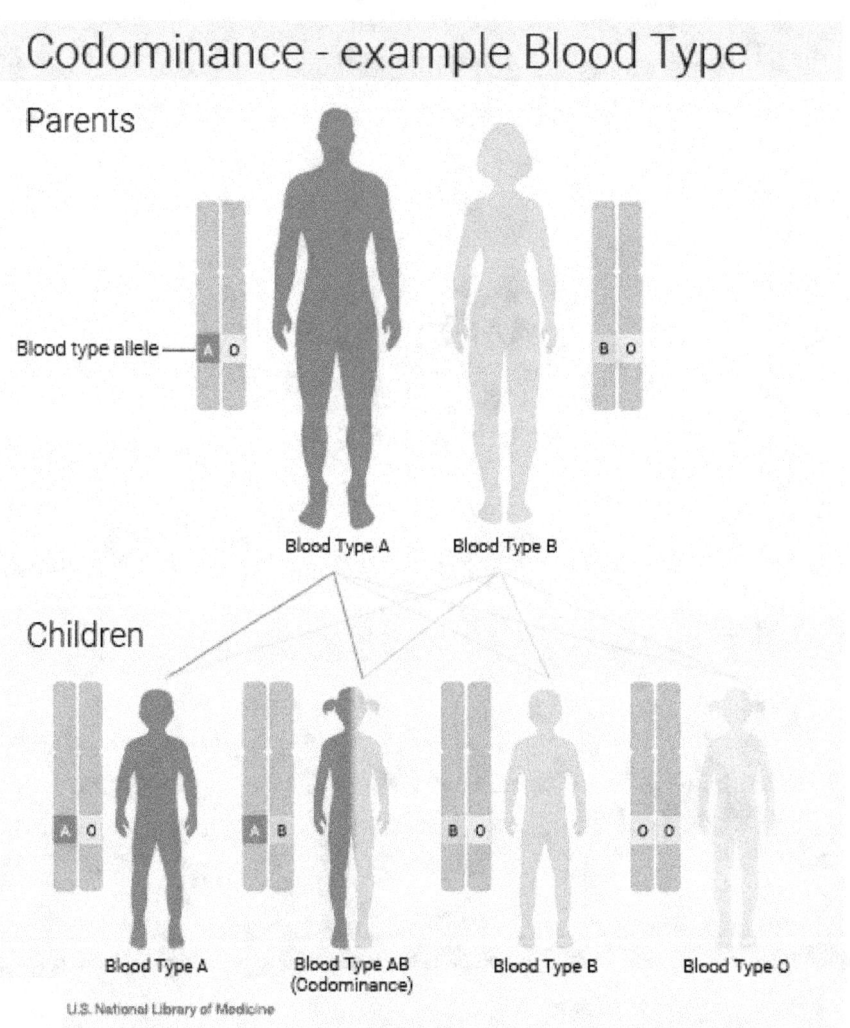

The ABO blood group is a major system for classifying blood types in humans. Blood type AB is inherited in a codominant pattern. In this example, a father with blood type A and a mother with blood type B have four children, each with a different blood type: A, AB, B, and O.

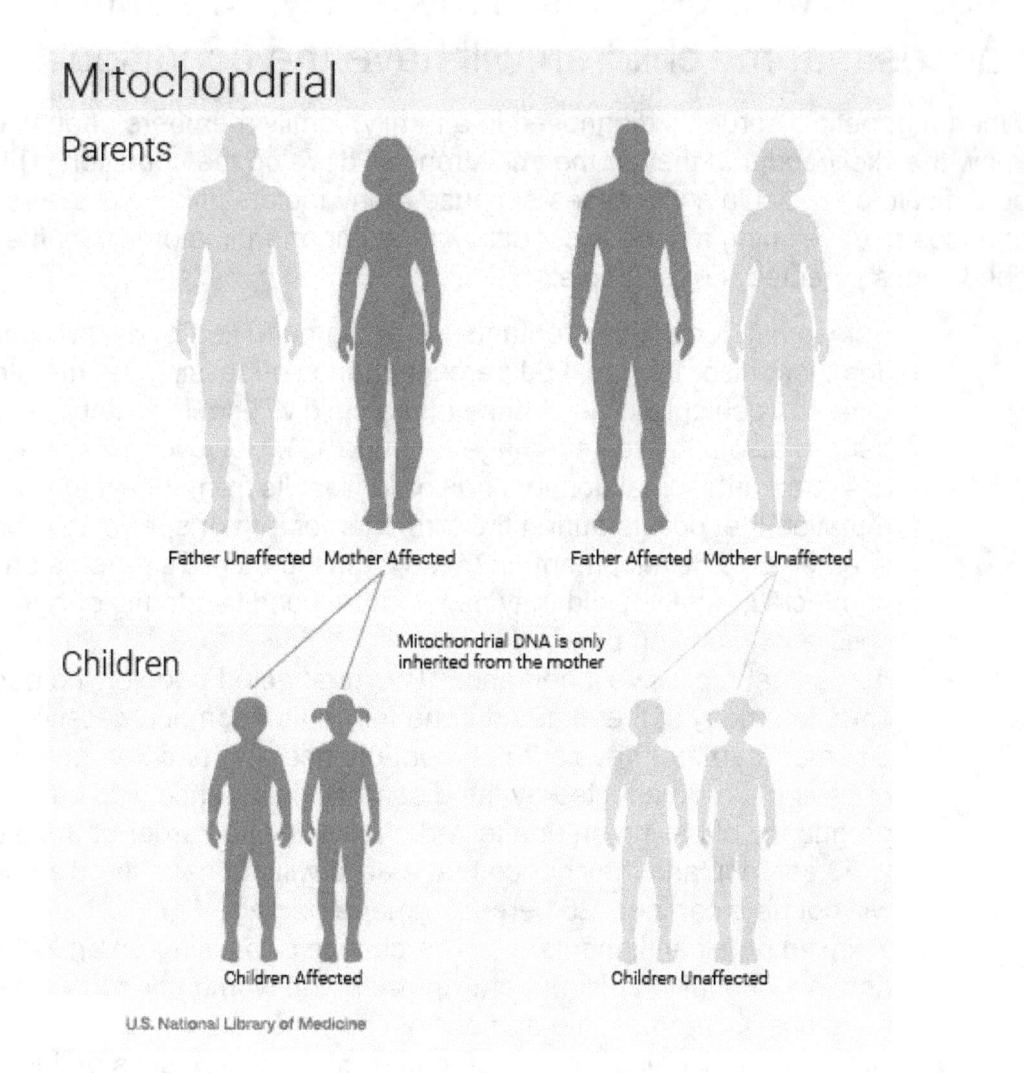

In the family on the left, a woman with a disorder caused by a mutation in mitochondrial DNA and her unaffected husband have children who are all affected by the condition. In the family on the right, a man with a condition resulting from a mutation in mitochondrial DNA and his unaffected wife have no affected children.

If a genetic disorder runs in my family, what are the chances that my children will have the condition?

When a genetic disorder is diagnosed in a family, family members often want to know the likelihood that they or their children will develop the condition. This can be difficult to predict in some cases because many factors influence a person's chances of developing a genetic condition. One important factor is how the condition is inherited. For example:

- Autosomal dominant inheritance: A person affected by an autosomal dominant disorder has a 50 percent chance of passing the mutated gene to each child. The chance that a child will not inherit the mutated gene is also 50 percent (image on page 12). However, in some cases an autosomal dominant disorder results from a new (de novo) mutation that occurs during the formation of egg or sperm cells or early in embryonic development. In these cases, the child's parents are unaffected, but the child may pass on the condition to his or her own children (image on page 13).
- Autosomal recessive inheritance: Two unaffected people who each carry one copy of the mutated gene for an autosomal recessive disorder (carriers) have a 25 percent chance with each pregnancy of having a child affected by the disorder. The chance with each pregnancy of having an unaffected child who is a carrier of the disorder is 50 percent, and the chance that a child will not have the disorder and will not be a carrier is 25 percent (image on page 14).
- X-linked dominant inheritance: The chance of passing on an X-linked dominant condition differs between men and women because men have one X chromosome and one Y chromosome, while women have two X chromosomes (image on page 15). A man passes on his Y chromosome to all of his sons and his X chromosome to all of his daughters. Therefore, the sons of a man with an X-linked dominant disorder will not be affected, but all of his daughters will inherit the condition. A woman passes on one or the other of her X chromosomes to each child. Therefore, a woman with an X-linked dominant disorder has a 50 percent chance of having an affected daughter or son with each pregnancy.

- X-linked recessive inheritance: Because of the difference in sex chromosomes, the probability of passing on an X-linked recessive disorder also differs between men and women (image on page 16). The sons of a man with an X-linked recessive disorder will not be affected, and his daughters will carry one copy of the mutated gene. With each pregnancy, a woman who carries an X-linked recessive disorder has a 50 percent chance of having sons who are affected and a 50 percent chance of having daughters who carry one copy of the mutated gene.
- Y-linked inheritance: Because only males have a Y chromosome, only males can be affected by and pass on Y-linked disorders. All sons of a man with a Y-linked disorder will inherit the condition from their father (image on page 17).
- Codominant inheritance: In codominant inheritance, each parent contributes a different version of a particular gene, and both versions influence the resulting genetic trait. The chance of developing a genetic condition with codominant inheritance, and the characteristic features of that condition, depend on which versions of the gene are passed from parents to their child (image on page 18).
- Mitochondrial inheritance: Mitochondria, which are the energy-producing centers inside cells, each contain a small amount of DNA. Disorders with mitochondrial inheritance result from mutations in mitochondrial DNA. Although these disorders can affect both males and females, only females can pass mutations in mitochondrial DNA to their children. A woman with a disorder caused by changes in mitochondrial DNA will pass the mutation to all of her daughters and sons, but the children of a man with such a disorder will not inherit the mutation (image on page 19).

It is important to note that the chance of passing on a genetic condition applies equally to each pregnancy. For example, if a couple has a child with an autosomal recessive disorder, the chance of having another child with the disorder is still 25 percent (or 1 in 4). Having one child with a disorder does not "protect" future children from inheriting the condition. Conversely, having a child without the condition does not mean that future children will definitely be affected.

Although the chances of inheriting a genetic condition appear straightforward, factors such as a person's family history and the results of genetic testing can sometimes modify those chances. In addition, some people with a disease-causing mutation never develop any health problems or may experience only mild symptoms of the disorder. If a disease that runs in a family does not have a clear-cut inheritance pattern, predicting the likelihood that a person will develop the condition can be particularly difficult.

Estimating the chance of developing or passing on a genetic disorder can be complex. Genetics professionals can help people understand these chances and help them make informed decisions about their health.

For more information about passing on a genetic disorder in a family:

The National Library of Medicine MedlinePlus web site offers information about the chance of developing a genetic disorder on the basis of its inheritance pattern:

- Autosomal dominant (https://medlineplus.gov/ency/article/002049.htm)
- Autosomal recessive (https://medlineplus.gov/ency/article/002052.htm)
- X-linked dominant (https://medlineplus.gov/ency/article/002050.htm)
- X-linked recessive (https://medlineplus.gov/ency/article/002051.htm)

The Centre for Genetics Education (Australia) provides an explanation of mitochondrial inheritance (http://www.genetics.edu.au/Publications-and-Resources/Genetics-Fact-Sheets/Fact%20Sheet%2012).

The Muscular Dystrophy Association explains patterns and probabilities (https://www.mda.org/sites/default/files/publications/Facts_Genetics_P-210_1.pdf) of inheritance.

What are reduced penetrance and variable expressivity?

Reduced penetrance and variable expressivity are factors that influence the effects of particular genetic changes. These factors usually affect disorders that have an autosomal dominant pattern of inheritance, although they are occasionally seen in disorders with an autosomal recessive inheritance pattern.

Reduced penetrance

Penetrance refers to the proportion of people with a particular genetic change (such as a mutation in a specific gene) who exhibit signs and symptoms of a genetic disorder. If some people with the mutation do not develop features of the disorder, the condition is said to have reduced (or incomplete) penetrance. Reduced penetrance often occurs with familial cancer syndromes. For example, many people with a mutation in the *BRCA1* or *BRCA2* gene will develop cancer during their lifetime, but some people will not. Doctors cannot predict which people with these mutations will develop cancer or when the tumors will develop.

Reduced penetrance probably results from a combination of genetic, environmental, and lifestyle factors, many of which are unknown. This phenomenon can make it challenging for genetics professionals to interpret a person's family medical history and predict the risk of passing a genetic condition to future generations.

Variable expressivity

Although some genetic disorders exhibit little variation, most have signs and symptoms that differ among affected individuals. Variable expressivity refers to the range of signs and symptoms that can occur in different people with the same genetic condition. For example, the features of Marfan syndrome vary widely— some people have only mild symptoms (such as being tall and thin with long, slender fingers), while others also experience life-threatening complications involving the heart and blood vessels. Although the features are highly variable, most people with this disorder have a mutation in the same gene (*FBN1*).

As with reduced penetrance, variable expressivity is probably caused by a combination of genetic, environmental, and lifestyle factors, most of which have not been identified. If a genetic condition has highly variable signs and symptoms, it may be challenging to diagnose.

For more information about reduced penetrance and variable expressivity:

The PHG Foundation offers an interactive tutorial on penetrance (http://www.phgfoundation.org/tutorials/penetrance/index.html) that explains the differences between reduced penetrance and variable expressivity.

Additional information about penetrance and expressivity (http://www.merckmanuals.com/home/fundamentals/genetics/inheritance-of-single-gene-disorders) is available from the Merck Manual Home Health Handbook for Patients & Caregivers.

What do geneticists mean by anticipation?

The signs and symptoms of some genetic conditions tend to become more severe and appear at an earlier age as the disorder is passed from one generation to the next. This phenomenon is called anticipation. Anticipation is most often seen with certain genetic disorders of the nervous system, such as Huntington disease, myotonic dystrophy, and fragile X syndrome.

Anticipation typically occurs with disorders that are caused by an unusual type of mutation called a trinucleotide repeat expansion. A trinucleotide repeat is a sequence of three DNA building blocks (nucleotides) that is repeated a number of times in a row. DNA segments with an abnormal number of these repeats are unstable and prone to errors during cell division. The number of repeats can change as the gene is passed from parent to child. If the number of repeats increases, it is known as a trinucleotide repeat expansion. In some cases, the trinucleotide repeat may expand until the gene stops functioning normally. This expansion causes the features of some disorders to become more severe with each successive generation.

Most genetic disorders have signs and symptoms that differ among affected individuals, including affected people in the same family. Not all of these differences can be explained by anticipation. A combination of genetic, environmental, and lifestyle factors is probably responsible for the variability, although many of these factors have not been identified. Researchers study multiple generations of affected family members and consider the genetic cause of a disorder before determining that it shows anticipation.

For more information about anticipation:

The Merck Manual for Healthcare Professionals provides a brief explanation of anticipation as part of its chapter on nontraditional inheritance (http://www.merckmanuals.com/professional/special-subjects/general-principles-of-medical-genetics/unusual-aspects-of-inheritance).

The Myotonic Dystrophy Foundation describes anticipation in the context of myotonic dystrophy (http://www.myotonic.org/what-dm/disease-mechanism). (Click on the tab that says "Anticipation.")

What are genomic imprinting and uniparental disomy?

Genomic imprinting and uniparental disomy are factors that influence how some genetic conditions are inherited.

Genomic imprinting

People inherit two copies of their genes—one from their mother and one from their father. Usually both copies of each gene are active, or "turned on," in cells. In some cases, however, only one of the two copies is normally turned on. Which copy is active depends on the parent of origin: some genes are normally active only when they are inherited from a person's father; others are active only when inherited from a person's mother. This phenomenon is known as genomic imprinting.

In genes that undergo genomic imprinting, the parent of origin is often marked, or "stamped," on the gene during the formation of egg and sperm cells. This stamping process, called methylation, is a chemical reaction that attaches small molecules called methyl groups to certain segments of DNA. These molecules identify which copy of a gene was inherited from the mother and which was inherited from the father. The addition and removal of methyl groups can be used to control the activity of genes.

Only a small percentage of all human genes undergo genomic imprinting. Researchers are not yet certain why some genes are imprinted and others are not. They do know that imprinted genes tend to cluster together in the same regions of chromosomes. Two major clusters of imprinted genes have been identified in humans, one on the short (p) arm of chromosome 11 (at position 11p15) and another on the long (q) arm of chromosome 15 (in the region 15q11 to 15q13).

Uniparental disomy

Uniparental disomy (UPD) occurs when a person receives two copies of a chromosome, or part of a chromosome, from one parent and no copies from the other parent. UPD can occur as a random event during the formation of egg or sperm cells or may happen in early fetal development.

In many cases, UPD likely has no effect on health or development. Because most genes are not imprinted, it doesn't matter if a person inherits both copies from one parent instead of one copy from each parent. In some cases, however, it does make a difference whether a gene is inherited from a person's mother or father. A person with UPD may lack any active copies of essential genes that undergo genomic imprinting. This loss of gene function can lead to delayed development, intellectual disability, or other health problems.

Several genetic disorders can result from UPD or a disruption of normal genomic imprinting. The most well-known conditions include Prader-Willi syndrome, which is characterized by uncontrolled eating and obesity, and Angelman syndrome, which causes intellectual disability and impaired speech. Both of these disorders can be caused by UPD or other errors in imprinting involving genes on the long arm of chromosome 15. Other conditions, such as Beckwith-Wiedemann syndrome (a disorder characterized by accelerated growth and an increased risk of cancerous tumors), are associated with abnormalities of imprinted genes on the short arm of chromosome 11.

For more information about genomic imprinting and UPD:

The University of Utah offers a basic overview of genomic imprinting (http://learn.genetics.utah.edu/content/epigenetics/imprinting/).

Additional information about epigenetics, including genomic imprinting (http://www.genetics.edu.au/Publications-and-Resources/Genetics-Fact-Sheets/FactSheetEpigenetics) is available from the Centre for Genetics Education.

An animated tutorial from the University of Miami illustrates how uniparental disomy occurs (http://hihg.med.miami.edu/code/http/modules/education/Design/animate/uniDisomy.htm).

Are chromosomal disorders inherited?

Although it is possible to inherit some types of chromosomal abnormalities, most chromosomal disorders (such as Down syndrome and Turner syndrome) are not passed from one generation to the next.

Some chromosomal conditions are caused by changes in the number of chromosomes. These changes are not inherited, but occur as random events during the formation of reproductive cells (eggs and sperm). An error in cell division called nondisjunction results in reproductive cells with an abnormal number of chromosomes. For example, a reproductive cell may accidentally gain or lose one copy of a chromosome. If one of these atypical reproductive cells contributes to the genetic makeup of a child, the child will have an extra or missing chromosome in each of the body's cells.

Changes in chromosome structure can also cause chromosomal disorders. Some changes in chromosome structure can be inherited, while others occur as random accidents during the formation of reproductive cells or in early fetal development. Because the inheritance of these changes can be complex, people concerned about this type of chromosomal abnormality may want to talk with a genetics professional.

Some cancer cells also have changes in the number or structure of their chromosomes. Because these changes occur in somatic cells (cells other than eggs and sperm), they cannot be passed from one generation to the next.

For more information about how chromosomal changes occur:

As part of its fact sheet on chromosome abnormalities, the National Human Genome Research Institute provides a discussion of how chromosome abnormalities happen. (https://www.genome.gov/11508982#6)

The Chromosome Deletion Outreach fact sheet Introduction to Chromosomes (http://chromodisorder.org/Display.aspx?ID=35) explains how structural changes occur.

The March of Dimes discusses the causes of chromosomal abnormalities in their fact sheet Chromosomal Abnormalities (http://www.marchofdimes.org/baby/chromosomal-conditions.aspx).

Additional information about how chromosomal changes happen (https://www.urmc.rochester.edu/Encyclopedia/Content.aspx?ContentTypeID=90&ContentID=P02126) is available from the University of Rochester Medical Center.

Why are some genetic conditions more common in particular ethnic groups?

Some genetic disorders are more likely to occur among people who trace their ancestry to a particular geographic area. People in an ethnic group often share certain versions of their genes, which have been passed down from common ancestors. If one of these shared genes contains a disease-causing mutation, a particular genetic disorder may be more frequently seen in the group.

Examples of genetic conditions that are more common in particular ethnic groups are sickle cell anemia, which is more common in people of African, African American, or Mediterranean heritage; and Tay-Sachs disease, which is more likely to occur among people of Ashkenazi (eastern and central European) Jewish or French Canadian ancestry. It is important to note, however, that these disorders can occur in any ethnic group.

For more information about genetic disorders that are more common in certain groups:

The National Coalition for Health Professional Education in Genetics offers Some Frequently Asked Questions and Answers About Race, Genetics, and Healthcare (http://www.nchpeg.org/index.php?option=com_content&view=article&id=142&Itemid=64).

The Norton & Elaine Sarnoff Center for Jewish Genetics provides information on disorders that occur more frequently in people with Jewish ancestry, including genetic traits that tend to be more common in Ashkenazi Jews (http://www.jewishgenetics.org/cjg/Ashkenazi-Jewish-Disorders.aspx) and Sephardic Jews (http://www.jewishgenetics.org/cjg/Sephardic-Jewish-Disorders.aspx).

https://ghr.nlm.nih.gov/

Lister Hill National Center for Biomedical Communications
U.S. National Library of Medicine
National Institutes of Health
Department of Health & Human Services

Published on January 17, 2017

www.ingramcontent.com/pod-product-compliance
Lightning Source LLC
Chambersburg PA
CBHW081316180526
45170CB00007B/2736